WHERE DO WORMS GO IN WINTER?

ANSWERING KIDS' QUESTIONS

by Ellen Labrecque

PEBBLE

a capstone imprint

Pebble Emerge is published by Pebble, an imprint of Capstone.
1710 Roe Crest Drive
North Mankato, Minnesota 56003
www.capstonepub.com

Library of Congress Cataloging-in-Publication Data
Names: Labrecque, Ellen, author.
Title: Where do worms go in winter? : answering kids' questions / by Ellen Labrecque.
Description: North Mankato, MN : Pebble, [2021] | Series: Questions and answers about animals | Includes bibliographical references and index. | Audience: Ages 6-8 | Audience: Grades 2-3 | Summary: "Birds have wings to fly somewhere warm in winter. But worms don't have wings or even any legs. Where do they go? You will learn all about worms and how they survive when the weather turns cold"—Provided by publisher.
Identifiers: LCCN 2020036422 (print) | LCCN 2020036423 (ebook) | ISBN 9781977131676 (hardcover) | ISBN 9781977132741 (paperback) | ISBN 9781977155337 (pdf) | ISBN 9781977156952 (kindle edition)
Subjects: LCSH: Worms—Wintering—Juvenile literature. | Animals—Wintering—Juvenile literature.
Classification: LCC QL386.6 .L33 2021 (print) | LCC QL386.6 (ebook) | DDC 592/.3—dc23
LC record available at https://lccn.loc.gov/2020036422
LC ebook record available at https://lccn.loc.gov/2020036423

Image Credits
Getty Images: Stan Tekiela Author/Naturalist/Wildlife Photographer, 11; iStockphoto: Christian Dahlhaus, 10, temmuzcan, 17; Newscom: Pete Oxford/Minden Pictures, 7; Science Source: Jacana, 15; Shutterstock: D. Kucharski K. Kucharska, 9, I like to take pictures, 13, kirillv alexey, design element, Kokhanchikov, 5, Nikolay Antonoc, Cover, oliveromg, 6, owatta, design element, SomprasongWittayanupakorn, 20, Witaya Proadtayakogool, 19

Editorial Credits
Editor: Megan Peterson; Designer: Ted Williams; Media Researcher: Jo Miller; Production Specialist: Spencer Rosio

All internet sites appearing in back matter were available and accurate when this book was sent to press.

Table of Contents

Words in **bold** are in the glossary.

Winter for Worms

Brr! The air is chilly. Soil freezes. Snow and ice cover the ground. Winter is here! Many earthworms live in gardens. But gardens bloom in summer. Worms can't crawl to warmer weather. They are too slow. They can't fly away either. What do worms do when winter comes?

Worm World

Thousands of earthworm **species** live almost everywhere on Earth. Many call warm, wet soil home. Some live in rotting logs or leaf piles.

Worms are **invertebrates**. They don't have backbones, eyes, or legs. Most have red-brown bodies. Some are blue or green! Many grow only a few inches long. One earthworm in South Africa grew to 22 feet (6.7 meters) long!

Living in Dirt

Earthworms love the dirt. Their bodies are built to live there. Worm bodies have many rings. Each ring is a muscle. The rings also have small, stiff hairs.

To move, the earthworm stretches out its muscles. The hairs grab dirt around the worm. Then the muscles grow shorter. They move the worm forward.

Mealtime

Earthworms **burrow** in the dirt. As they dig, they eat the dirt. The dirt has bits of dead plants and animals in it. The worms **digest** these bits. Worm poop passes out the back of the worm.

Worm poop is also called castings. Castings have lots of **nutrients**. They help plants grow. Worm tunnels also help the soil get water and air.

castings

Breathing Without Lungs

Earthworms don't have noses or lungs. They breathe through their skin. Their bodies make slime. The slime helps **oxygen** pass through their skin.

An earthworm needs wet skin to breathe. Dirt also helps keep the skin wet. If its skin dries out, the worm dies. But what happens when the soil freezes? Where do worms go?

Just Keep Digging

When winter comes, some worms dig deep into the dirt. They go below the frost line. The soil does not freeze below this line. Worms stay safe and warm there. The dirt acts like a cozy blanket. The worms curl up and sleep until spring.

Worm Babies

Other earthworms only live in the **topsoil**. They never go deep into the soil. When winter comes, these worms freeze. They lay eggs before they die.

The eggs stay in a sack during the winter. The sack keeps the eggs warm. They don't freeze. When spring comes, the young **hatch**. New earthworms arrive!

The Wonder of Worms

Earthworms are all around us. They help flowers and other plants grow. In winter, some worms sleep. These worms can live up to eight years.

Other earthworms freeze in winter. They live about one year. Before they die, they lay eggs. They keep their species alive season after season.

Grow with Worms

What You Need:

- two pots
- small shovel
- potting soil
- flower seeds
- water

What You Do:

1. Fill two pots with potting soil.

2. Have an adult help you plant flower seeds in the pots. Make sure each pot has the same number of seeds.

3. Carefully dig up some worms from dirt outside.

4. Place one or two worms in one of the pots. Do not put any worms in the other pot.

5. Water your flowers each day.

6. See which flowers grow faster. It should be the flowers with the worms. Worms help flowers bloom!

Glossary

burrow (BUHR-oh)—to dig a hole in the ground

digest (dy-GEST)—to break down food so it can be used by the body

hatch (HACH)—to break out of an egg

invertebrate (in-VUR-tuh-bruht)—an animal without a backbone

nutrient (NOO-tree-uhnt)—something that is needed by people, animals, and plants to stay healthy and strong

oxygen (OK-suh-juhn)—a colorless gas that people and animals breathe

species (SPEE-sheez)—a group of animals with similar features

topsoil (TAHP-soil)—the layer of soil in which plants grow

Read More

Peterson, Megan Cooley. *Wonderful Worms*. North Mankato, MN: Pebble, a Capstone Imprint, 2020.

Statts, Leo. *Earthworms*. Minneapolis: Abdo Zoom, 2017.

Williams, Susie. *Worms*. New York: Crabtree Publishing Company, 2020.

Internet Sites

Fact Monster: DK Science: Worms
factmonster.com/dk/encyclopedia/science/worms

National Geographic Kids: Earthworm
kids.nationalgeographic.com/animals/invertebrates/earthworm/

Natural History Museum: Earthworms
nhm.ac.uk/discover/earthworm-heroes.html

Index